Today was moving day at Theo and Gus's house.

Gus helped his mother wash the floors. Theo took down the last pictures.

Finally the moving van came.
Everyone was busy!

Theo and Gus knew they would miss their neighbors. They drove away with heavy hearts.

But the sight of their new house made them feel joyful.

The next morning, a friendly girl showed Theo and Gus around school.

Later, Miss Green asked Gus to talk about himself. Gus felt shy, but he felt special too.

Theo and Gus missed their old friends. But they no longer felt like strangers!